Comparing Quantities

Britannica

Mathematics in Context is a comprehensive curriculum for the middle grades. It was developed in collaboration with the Wisconsin Center for Education Research, School of Education, University of Wisconsin–Madison and the Freudenthal Institute at the University of Utrecht, The Netherlands, with the support of National Science Foundation Grant No. 9054928.

National Science Foundation

Opinions expressed are those of the authors
and not necessarily those of the Foundation

Revision Project

Peter Sickler
Project Director

Teri Hedges
Revision Consultant

Nieka Mamczak
Revision Consultant

Erin Turner
Revision Consultant

Cheryl Deese
MiC General Manager

Vicki Mirabile
Project Manager

ISBN 0-03-071516-4

2 3 4 5   170   05 04 03

The *Mathematics in Context* Development Team

Mathematics in Context is a comprehensive curriculum for the middle grades. The National Science Foundation funded the National Center for Research in Mathematical Sciences Education at the University of Wisconsin–Madison to develop and field-test the materials from 1991 through 1996. The Freudenthal Institute at the University of Utrecht in The Netherlands, as a subcontractor, collaborated with the University of Wisconsin–Madison on the development of the curriculum.

The initial version of *Comparing Quantities* was developed by Martin Kindt and Mieke Abels. It was adapted for use in American schools by Margaret R. Meyer and Margaret A. Pligge.

National Center for Research in Mathematical Sciences Education Staff

Thomas A. Romberg
Director

Joan Daniels Pedro
Assistant to the Director

Gail Burrill
Coordinator
Field Test Materials

Margaret R. Meyer
Coordinator
Pilot Test Materials

Mary Ann Fix
Editorial Coordinator

Sherian Foster
Editorial Coordinator

James A. Middleton
Pilot Test Coordinator

Project Staff

Jonathan Brendefur
Laura J. Brinker
James Browne
Jack Burrill
Rose Byrd
Peter Christiansen
Barbara Clarke
Doug Clarke
Beth R. Cole

Fae Dremock
Jasmina Milinkovic
Margaret A. Pligge
Mary C. Shafer
Julia A. Shew
Aaron N. Simon
Marvin Smith
Stephanie Z. Smith
Mary S. Spence

Freudenthal Institute Staff

Jan de Lange
Director

Els Feijs
Coordinator

Martin van Reeuwijk
Coordinator

Project Staff

Mieke Abels
Nina Boswinkel
Frans van Galen
Koeno Gravemeijer
Marja van den Heuvel-Panhuizen
Jan Auke de Jong
Vincent Jonker
Ronald Keijzer

Martin Kindt
Jansie Niehaus
Nanda Querelle
Anton Roodhardt
Leen Streefland
Adri Treffers
Monica Wijers
Astrid de Wild

Table of Contents

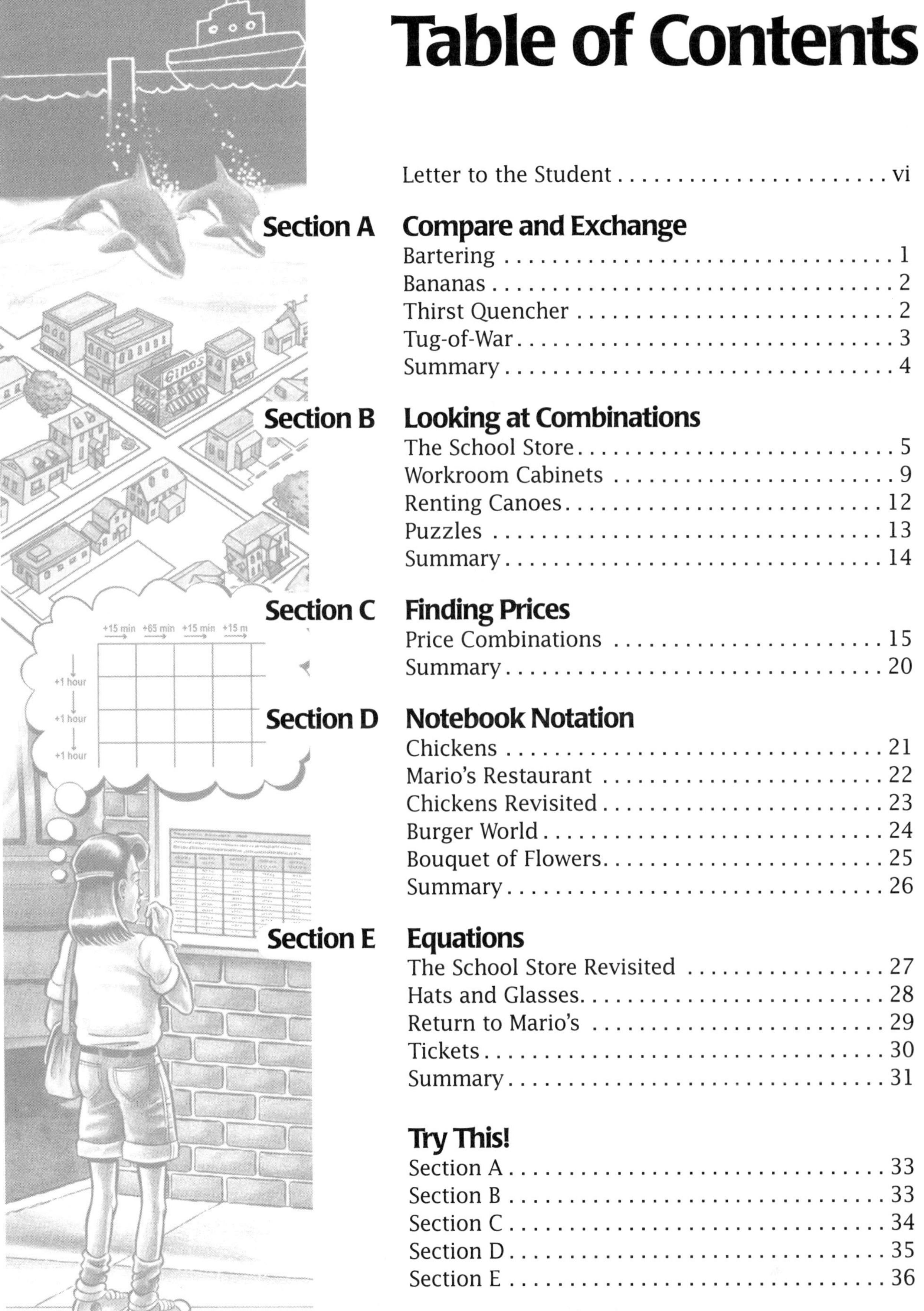

Dear Student,

Welcome to *Comparing Quantities.*

In this unit, you will compare quantities such as prices, weights, or widths.

You will learn about trading or exchanging things in order to find information such as how many bananas are needed to make the third scale balance.

You will use charts to help you plan a canoe trip. The number of canoes rented will depend on the number of people going on the trip and the size of the canoes.

In the end, you will have learned important ideas about algebra and several new ways to solve problems.

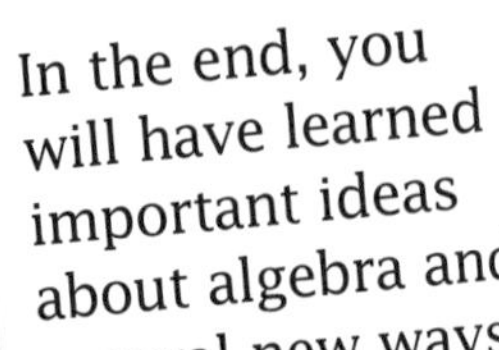
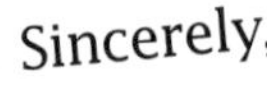

Sincerely,

The Mathematics in Context Development Team

Bartering

There was a time when money didn't exist. People lived in small communities, grew their own crops, and raised cattle or sheep. What did they do if they needed something they didn't produce themselves? They traded some of the things they produced for the things they needed. This is called *bartering* or exchanging.

Paulo lives with his family in a small village. His family needs corn. He is going to the market with two sheep and one goat to exchange them for bags of corn.

First he meets Aaron, who says, "I only trade salt for chickens. I will give you one bag of salt for every two chickens."

"I don't have any chickens," thinks Paulo, "so I can't trade with Aaron."

Later he meets Sarkis who tells him, "I will give you two bags of corn for every three bags of salt."

"That doesn't help me either," Paulo thinks.

Then he meets Ranee. She will trade six chickens for every goat, and she says, "My sister, Nina, is willing to give you six bags of salt for every sheep you have."

Paulo is getting confused. What can he do? He has to come home with just bags of corn, not with a goat or sheep or chickens or salt.

1. Show what Paulo can do.

Bananas

2. How many bananas are needed to make the third scale balance? Explain your reasoning.

3. How many carrots are needed to make the third scale balance? Explain your reasoning.

Thirst Quencher

4. How many cups of liquid can you pour out of one big bottle? Explain your reasoning.

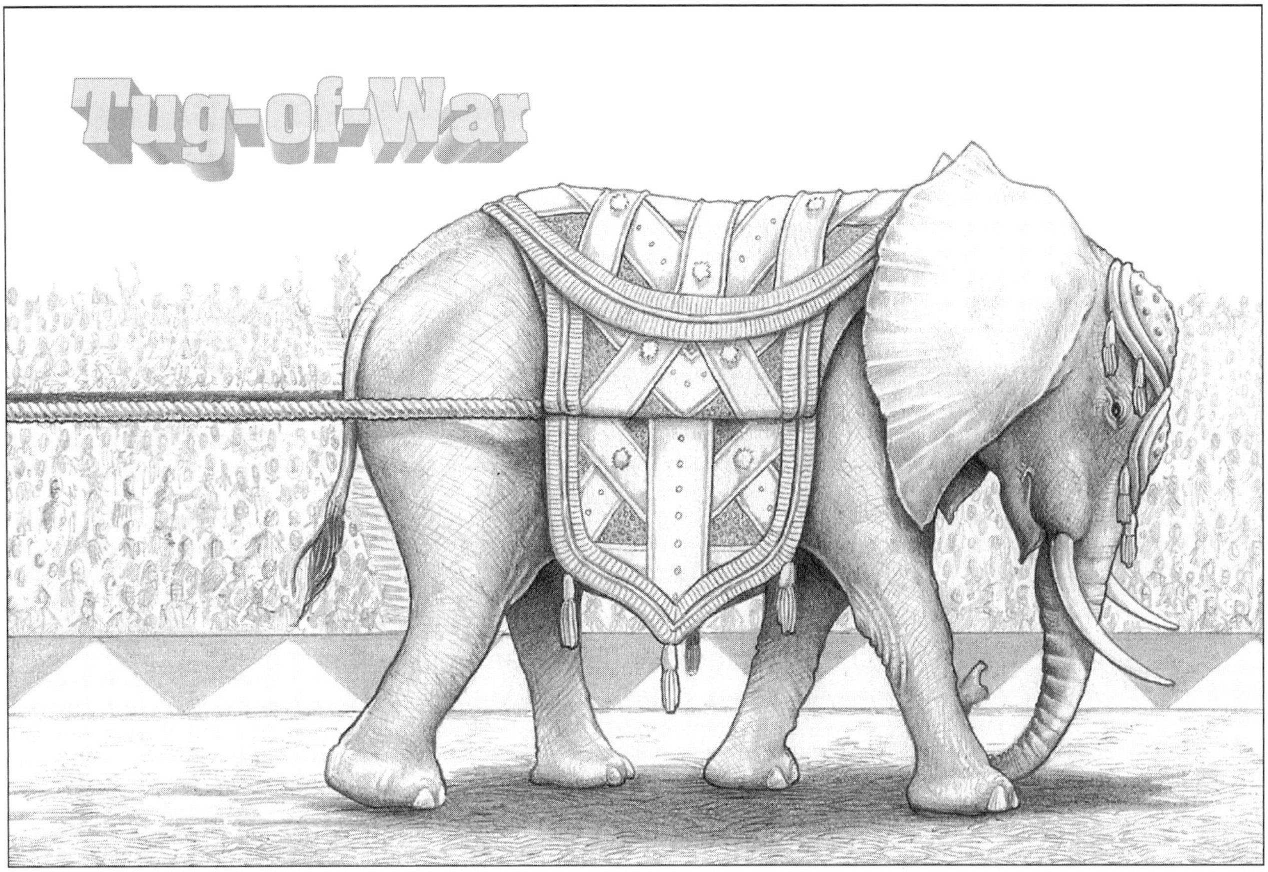

Four oxen are as strong as five horses.

An elephant is as strong as one ox and two horses.

5. Who will win the tug-of-war pictured below? Give a reason for your prediction.

Summary

Problems can be solved using the idea of exchanging. In this section, problems were given in words or pictures. You explained your work using words, pictures, or symbols.

Summary Questions

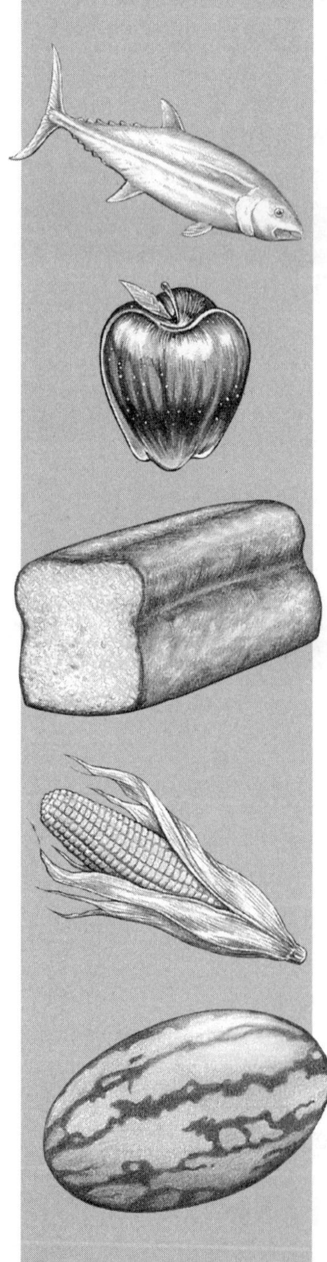

Delia lives in a community where the people trade goods they produce for other things they need. Delia has some fish that she caught, and she wants to trade them for other food. Delia hears that she can trade fish for melons. She wants more than just melons, so she decides to see what else is available. This is what she hears:

- For five fish you can get two melons.

- For four apples you can get one loaf of bread.

- For one melon you can get one ear of corn and two apples.

- For ten apples you can get four melons.

6. Rewrite or draw pictures of this information so that it is easier to use.

7. Use the statements above to make up two more statements about the exchanging of apples, melons, corn, fish, and bread.

8. Delia says, "I can get 10 apples for 10 fish." Is this true? Explain.

9. Is it also true that she can trade three fish for one loaf of bread? Explain why or why not.

10. Explain how Delia can get some ears of corn.

The School Store

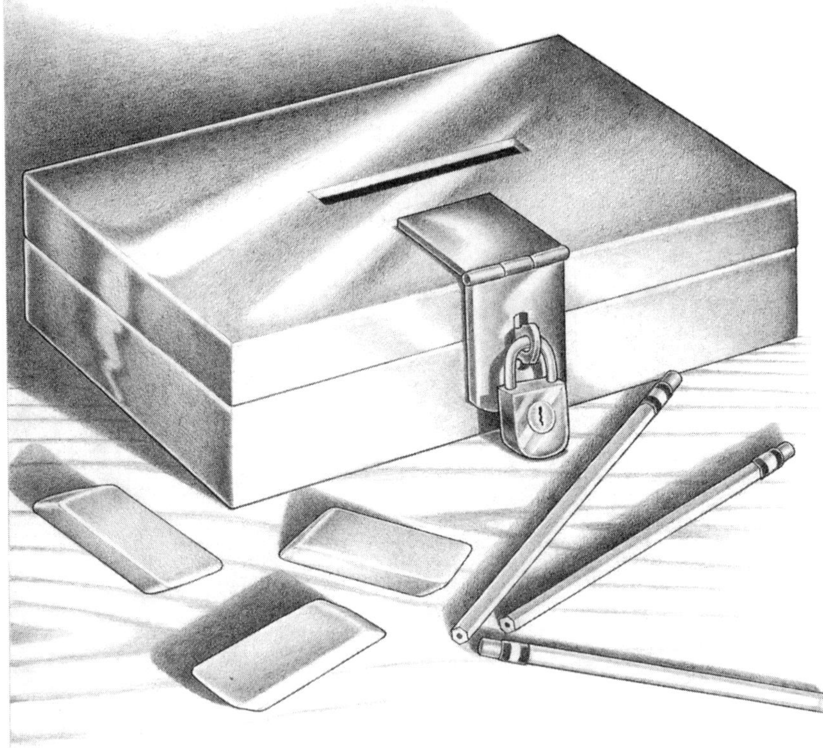

Monica and Martin are in charge of the school store. The store is open all day for students to come and buy supplies. Unfortunately, Monica and Martin can't be there all day to take students' money, so they start an honor system. Limited quantities of pencils and erasers are available for students to purchase using this honor system. Students leave the exact change in a small, locked box. Erasers cost 25¢ each, and pencils cost 15¢ each.

1. One day Monica and Martin find $1.10 in the locked box. How many pencils and how many erasers were purchased?

2. On another day there is $1.50 in the locked box. Monica and Martin cannot decide what was purchased. Why?

3. Find another amount of money for which you could not be sure what had been purchased.

Monica wants to find a way to make the calculations easier, so she makes two price lists: one for different numbers of erasers and one for different numbers of pencils.

4. Copy and complete the price lists for the erasers and the pencils.

Erasers	Price
0	$0.00
1	$0.25
2	$0.50
3	$0.75
4	$1.00
5	$1.25
6	
7	

Pencils	Price
0	$0.00
1	$0.15
2	$0.30
3	
4	
5	
6	
7	

One day the box has $1.05 in it.

5. Show how Monica can use her lists to find out how many pencils and erasers were bought.

Monica and Martin aren't satisfied. Although they now have these two lists, they still have to do many calculations. They are trying to think of a way to get all the prices for all the combinations of pencils and erasers in one chart.

6. What suggestions can you make for combining the two lists? Discuss your ideas with your class.

Combination Chart

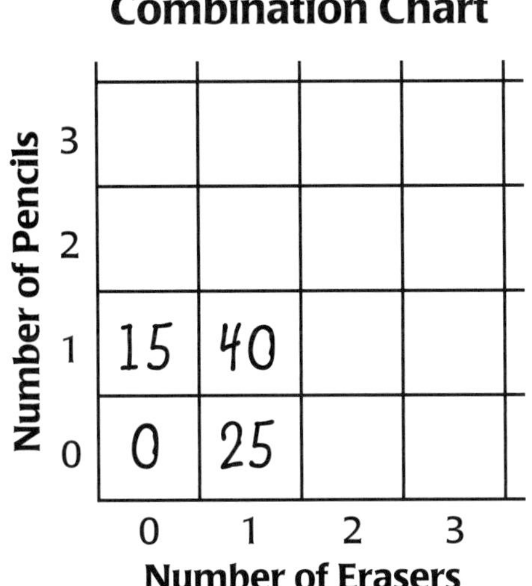

Monica and Martin come up with the idea of a *combination chart*. On the left you see what part of their chart looks like.

7. a. What does the 40 in the chart mean?

b. How many combinations of erasers and pencils can Monica and Martin put in this part of the chart?

If you extend this chart, as shown below, you can put in more combinations.

Use the combination chart on **Student Activity Sheet 1** to solve the following problems.

8. Fill in the white squares with the prices of the combinations.

9. Circle the price of two erasers and three pencils.

Costs of Combinations (in cents)

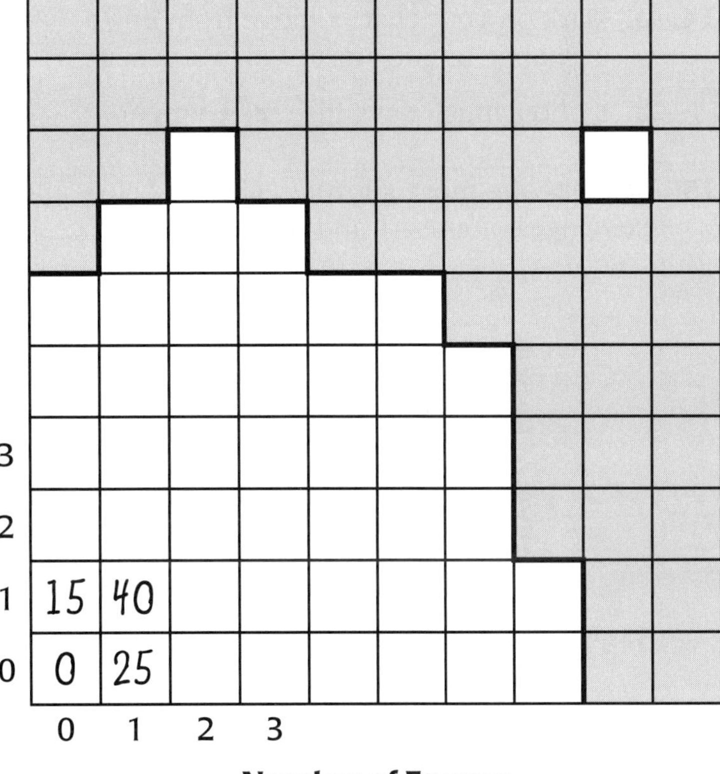

Use number patterns in your completed combination chart on **Student Activity Sheet 1** to answer problems **10–16**.

10. a. Where do you find the answer to problem **1** on page 5 (110¢) in the chart?

b. How many erasers and how many pencils were bought for this amount?

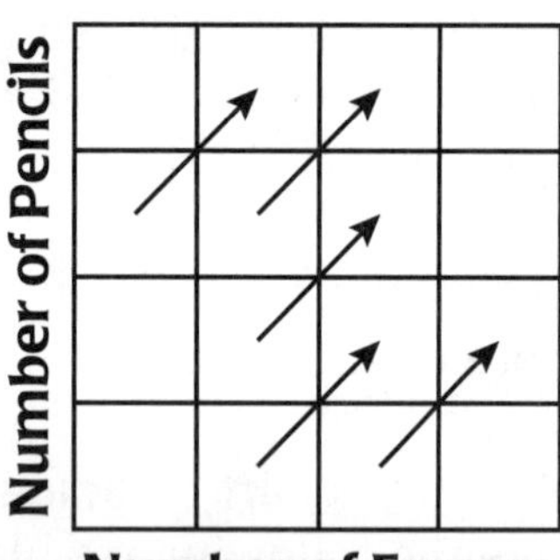

11. a. What happens to the numbers in the chart as you move along one of the arrows shown in the diagram on the left?

b. Is the answer to part **a** different depending on which arrow you choose?

12. What does moving along an arrow mean in terms of the numbers of pencils and erasers purchased?

13. a. Mark on your chart a move from one square to another that represents the exchange of one pencil for one eraser.

b. By how much does the cost change?

14. a. Mark on your chart a move from one square to another that represents the exchange of one eraser for two pencils.

b. By how much does the cost change?

15. Describe the move shown in chart **a** below and the one in chart **b** in terms of the exchanges of erasers and pencils.

a.

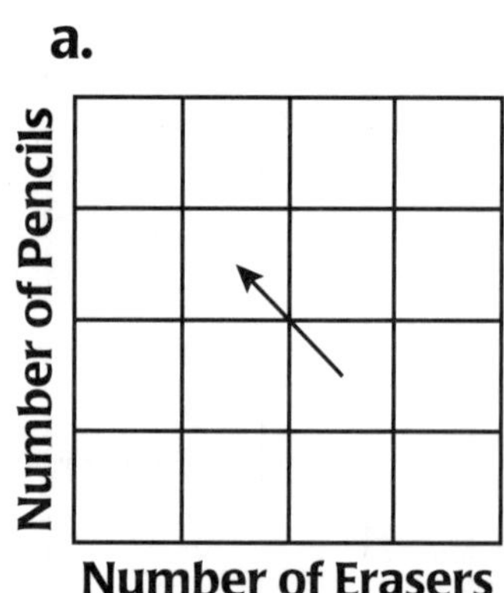

b.

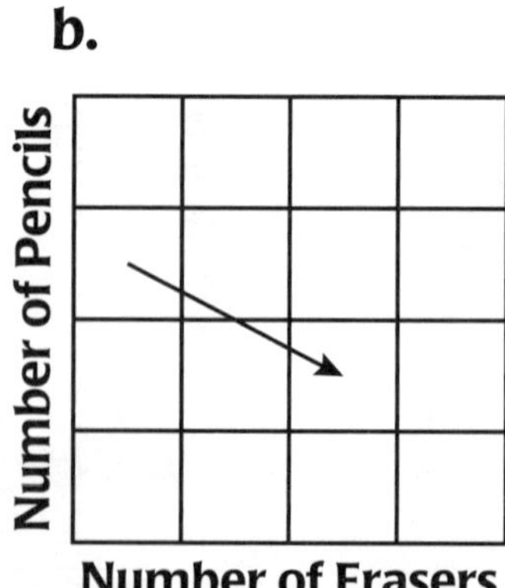

16. There are many other moves and patterns in the chart. Find at least two other patterns. Use different-colored pencils to mark them on your chart. Describe each pattern you find.

Workroom Cabinets

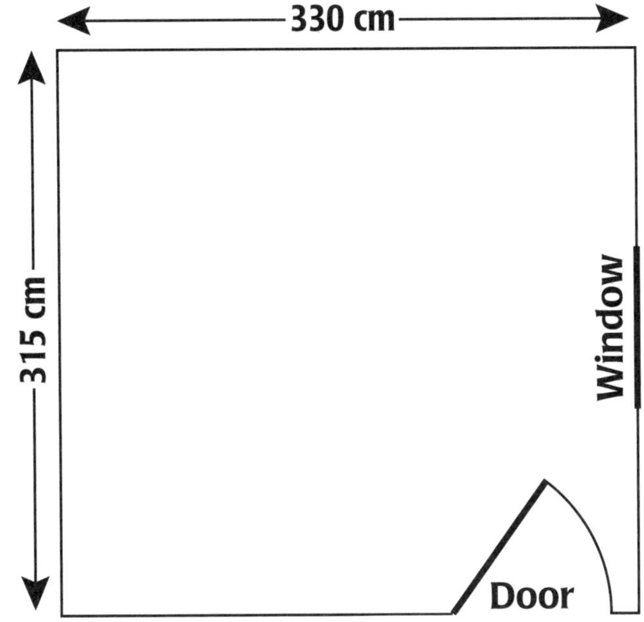

Ms. Simon is going to remodel her workroom. She wants to put new cabinets along one wall of the room. She starts by measuring the room and drawing the diagram shown on the left.

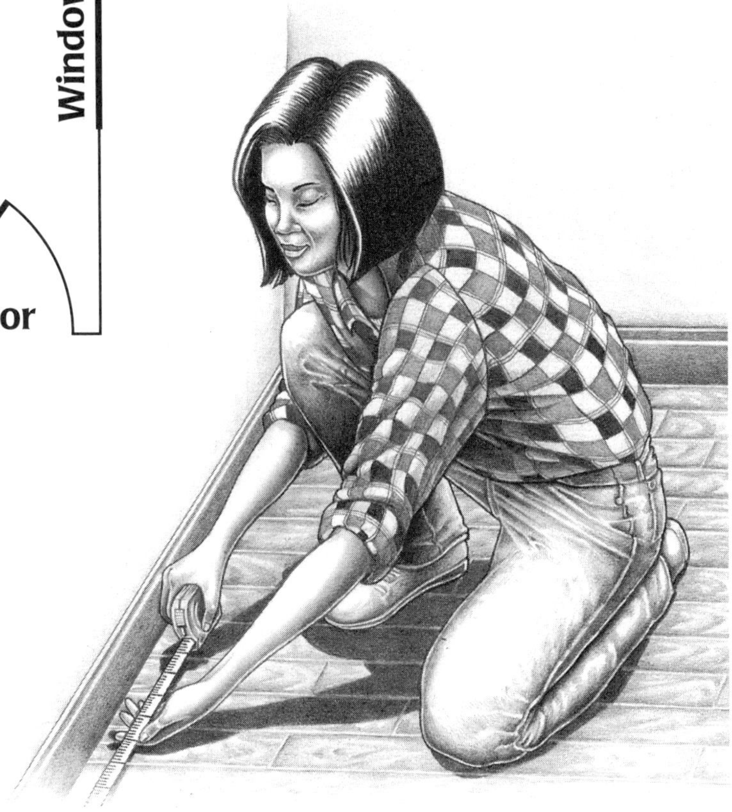

Ms. Simon finds out that the cabinets come in two different widths: 45 centimeters and 60 centimeters.

17. How many of each cabinet does Ms. Simon need in order for the cabinets to fit exactly along the wall that measures 315 centimeters? Try to find more than one possibility.

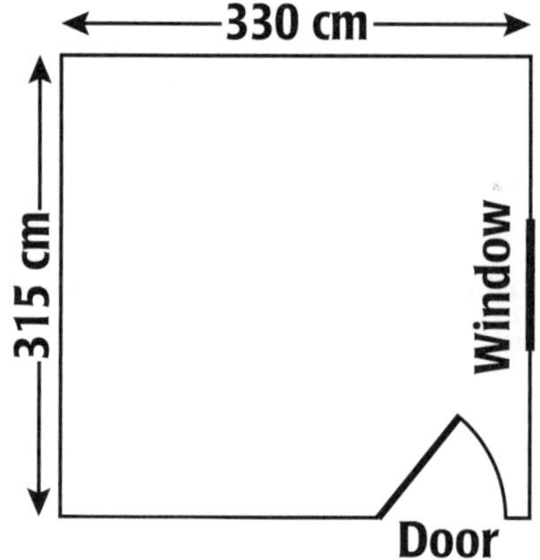

Ms. Simon wonders how she can design cabinets for the longer wall.

The cabinet store has a convenient chart. Using this chart makes it easy to find out how many 60-centimeter and 45-centimeter cabinets are needed for different wall lengths.

18. Explain how Ms. Simon can use the chart to find the number of cabinets she needs for the longer wall in her workroom.

Lengths of Combinations (in centimeters)

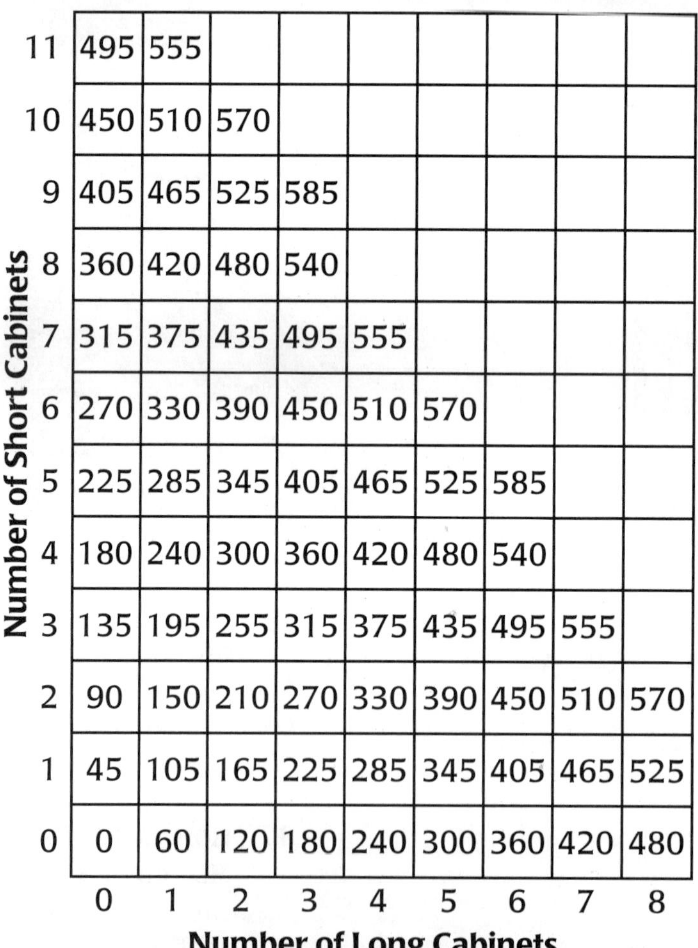

Number of Short Cabinets	0	1	2	3	4	5	6	7	8
11	495	555							
10	450	510	570						
9	405	465	525	585					
8	360	420	480	540					
7	315	375	435	495	555				
6	270	330	390	450	510	570			
5	225	285	345	405	465	525	585		
4	180	240	300	360	420	480	540		
3	135	195	255	315	375	435	495	555	
2	90	150	210	270	330	390	450	510	570
1	45	105	165	225	285	345	405	465	525
0	0	60	120	180	240	300	360	420	480

Number of Long Cabinets

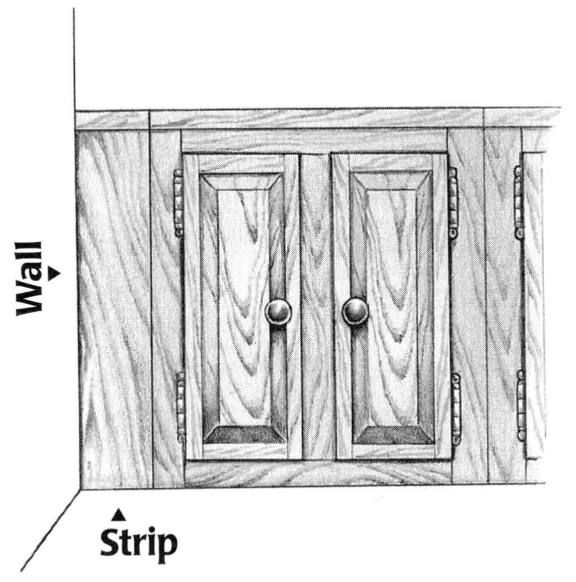

19. Can the cabinet store provide cabinets to fit exactly 4 meters? Explain your answer.

If cabinets don't fit exactly, the cabinet store sells a strip to fill the gap. Most customers want the strip to be as small as possible.

20. What size strip is necessary for cabinets along a 4-meter wall?

The chart has been completed only up to 585 centimeters because longer rows of cabinets are not often purchased. However, one day an order comes in for cabinets that fit a wall exactly 6 meters long. One possibility is, of course, 10 cabinets of 60 centimeters each.

21. What are other possibilities for a cabinet arrangement that will fit a 6-meter wall? Note that although you do not see 600 in the chart, you can still use the chart. How?

Below is a part of the cabinet combination chart.

Lengths of Combinations (in centimeters)

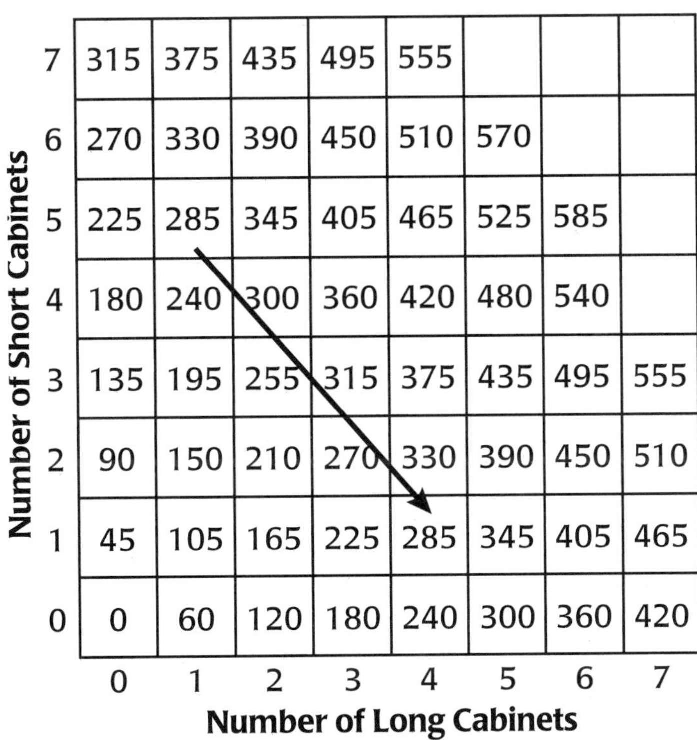

Number of Short Cabinets	0	1	2	3	4	5	6	7
7	315	375	435	495	555			
6	270	330	390	450	510	570		
5	225	285	345	405	465	525	585	
4	180	240	300	360	420	480	540	
3	135	195	255	315	375	435	495	555
2	90	150	210	270	330	390	450	510
1	45	105	165	225	285	345	405	465
0	0	60	120	180	240	300	360	420

Number of Long Cabinets

22. What is special about the move shown by the arrow?

23. If you start in another square in this chart and you make the same move, what do you notice? How can you explain this?

Renting Canoes

A Girl Scout troop wants to rent canoes for a group of 25 people. There are small and large canoes available. Each small canoe carries two people, and each large canoe carries three people.

Use the combination chart on **Student Activity Sheet 2** to solve the following problems. You do not have to fill in the whole chart.

24. What combinations of small and large canoes will accommodate exactly 25 people? Find all the different possibilities.

25. One person broke her leg a week before the trip and is unable to go on the outing. What are the possibilities for the new reservation?

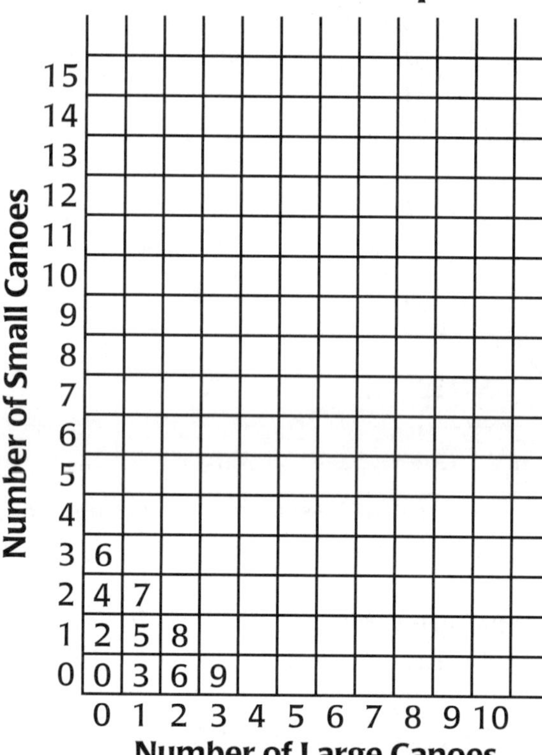

Numbers of People on Canoe Trip

Puzzles

a.

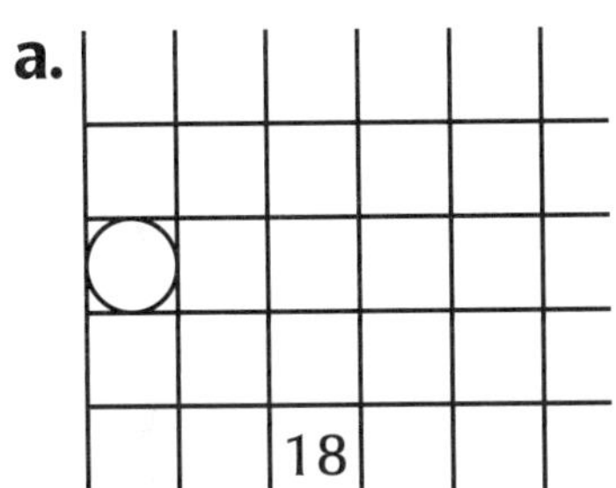

○					
		18			
0	5				

b.

	37				
	27				
0			○		

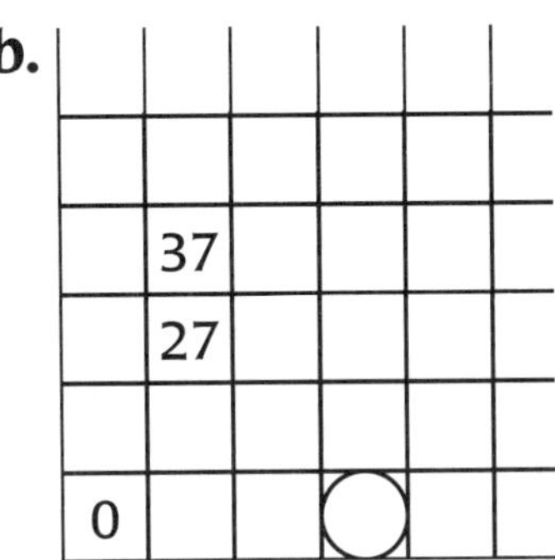

26. Complete the puzzles on **Student Activity Sheet 3.**

27. Make up your own puzzle chart. Give your puzzle to another student and have him or her solve it.

c.

		20		○	
			24		
0					

d.

				○	
	35				
			55		
0					

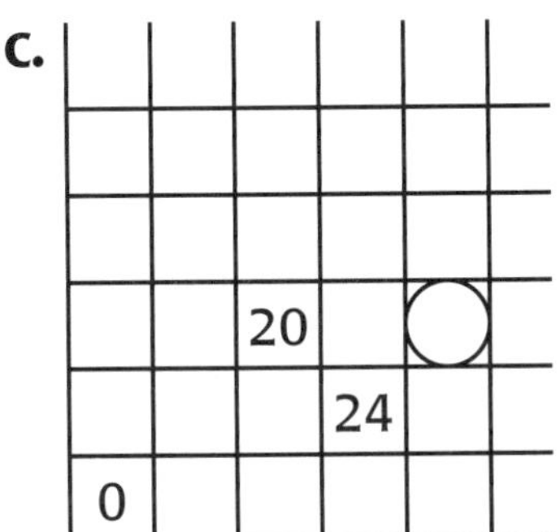

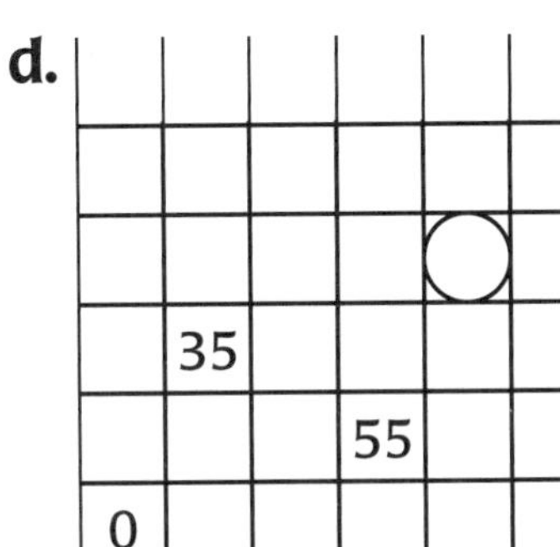

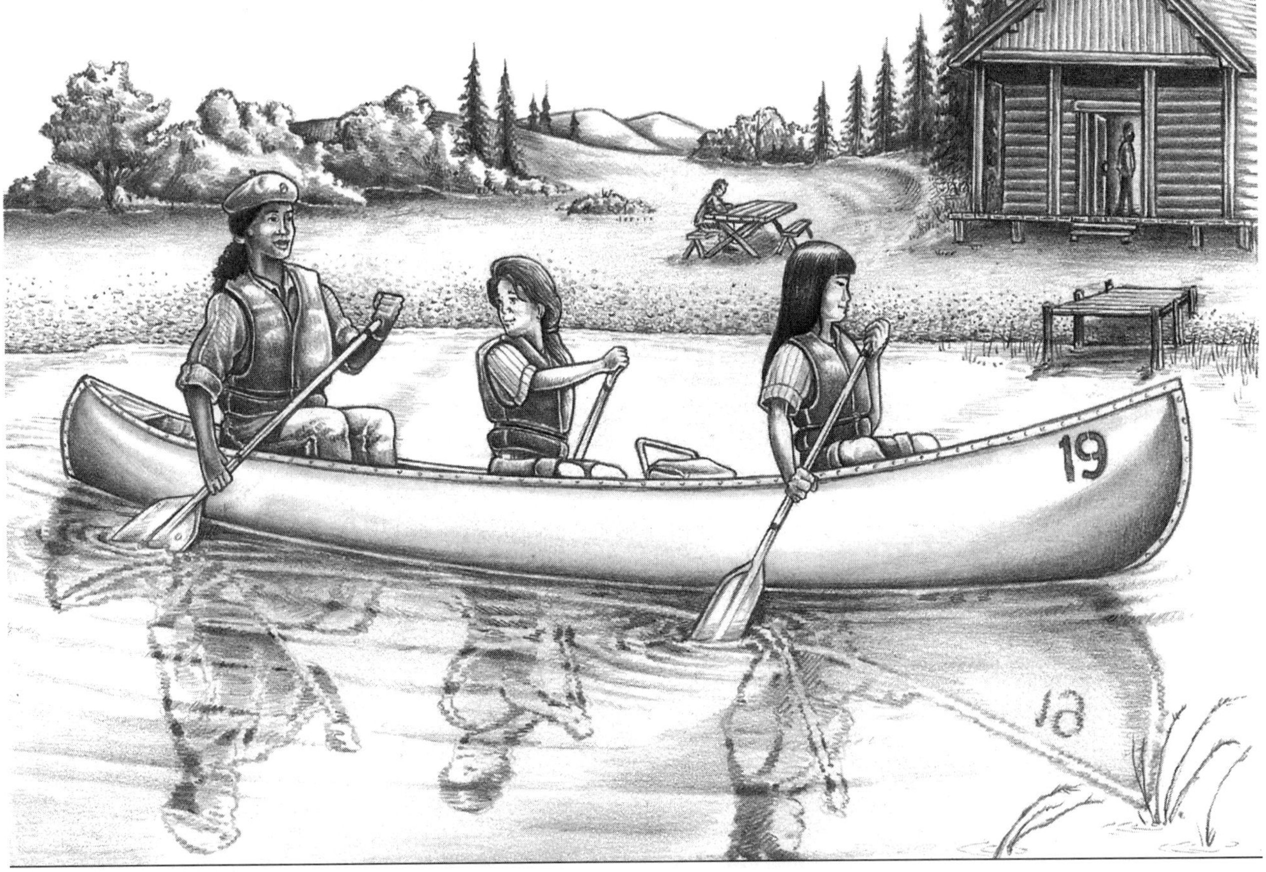

Summary

Combination charts show a quick view of many combinations. As you saw in "The School Store," "Workroom Cabinets," and "Renting Canoes," charts can be used to solve many different problems. Discovering patterns within combination charts can make your work easier.

Summary Questions

28. Write a story problem in your journal for which the combination chart below can be used. What do the circled numbers mean in your story?

50	52	54	56	58	60
(40)	42	44	46	48	50
30	32	34	36	38	40
20	22	24	26	28	30
10	12	14	(16)	18	20
0	2	4	6	8	10

Price Combinations

So far you have used two strategies for solving problems that involve combinations of items. The first strategy, exchanging, was used in the problems about trading food at the beginning of the unit. The second strategy was to make a combination chart and use number patterns found in the chart.

In this section, you will use the strategy of exchanging to solve problems involving money.

1. Without knowing the prices of a pair of glasses or a pair of shorts, you can determine which item is more expensive. Explain how.

2. How many pairs of shorts can you buy for $50?

3. What is the price of one pair of glasses? Explain your reasoning.

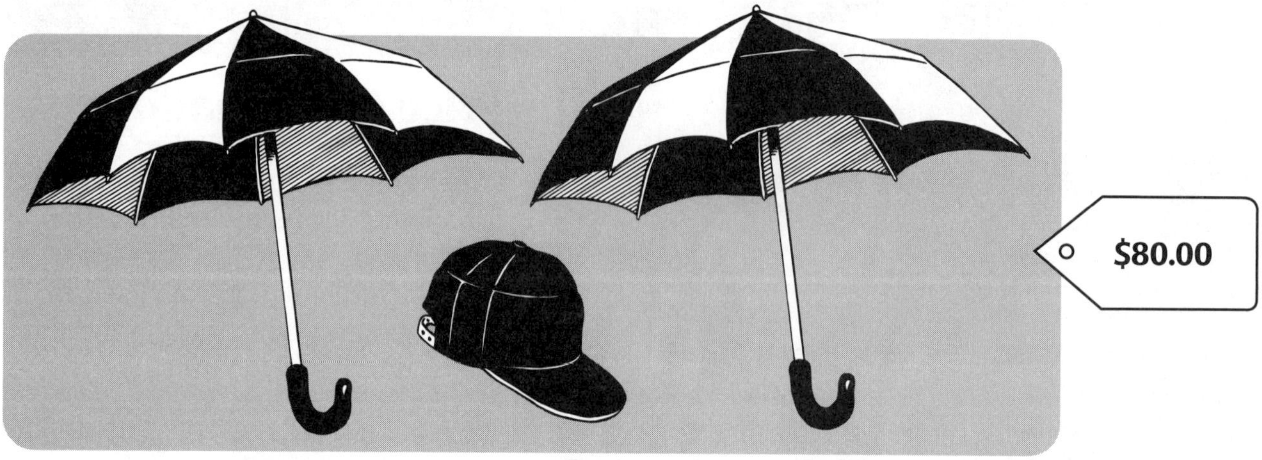

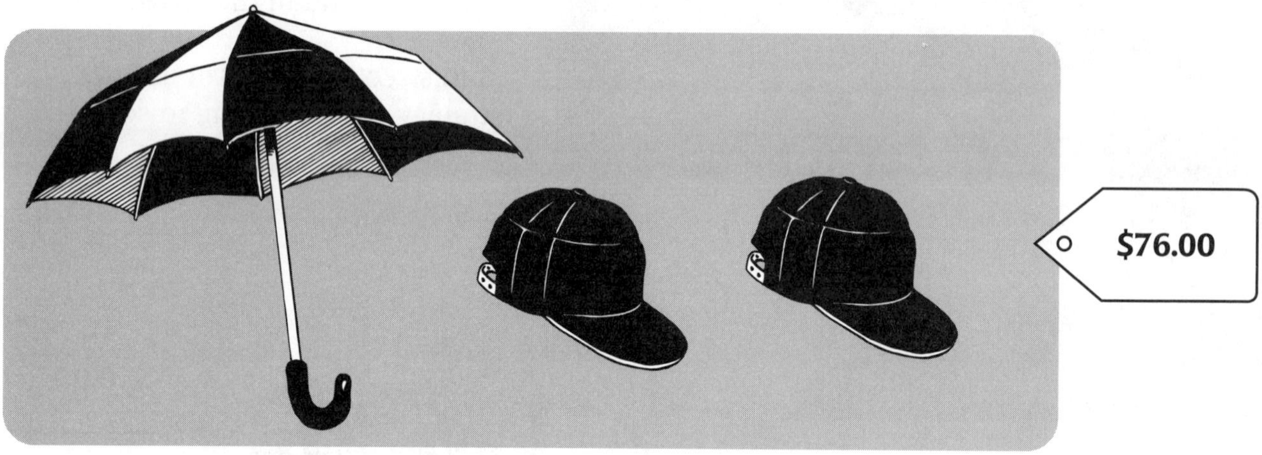

4. Which is more expensive, a cap or an umbrella? How much more expensive is it?

5. Use the two pictures above to make a new combination of umbrellas and caps. Write down the cost of the combination.

6. Make a group of only caps or only umbrellas. Then find its price.

7. What is the price of one umbrella? one cap?

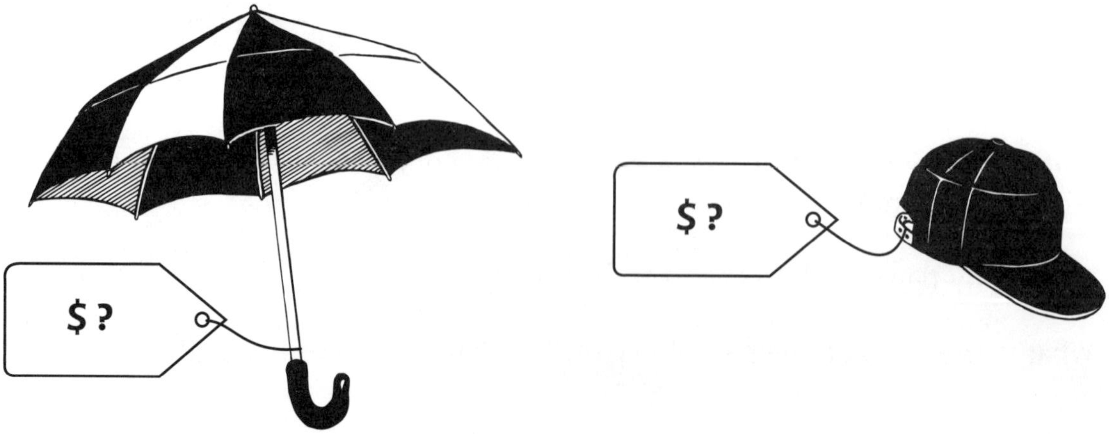

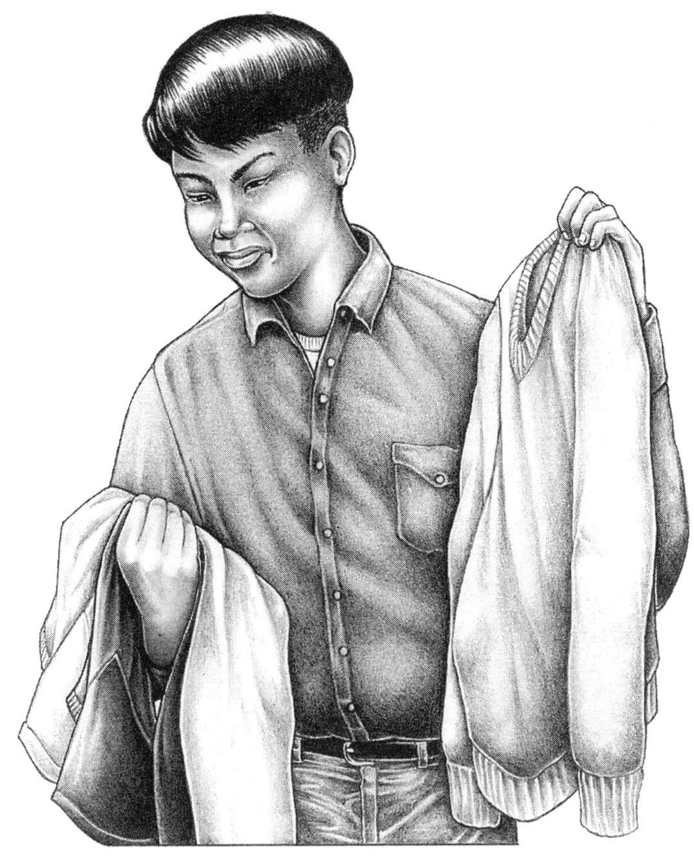

Sean bought two T-shirts and one sweatshirt for a total of $30. When he got home, he regretted his purchase. He decided to exchange one T-shirt for another sweatshirt.

Sean was able to do this, but he had to pay $6 more because the sweatshirt was more expensive than the T-shirt.

8. What was the price of each item? Explain your reasoning.

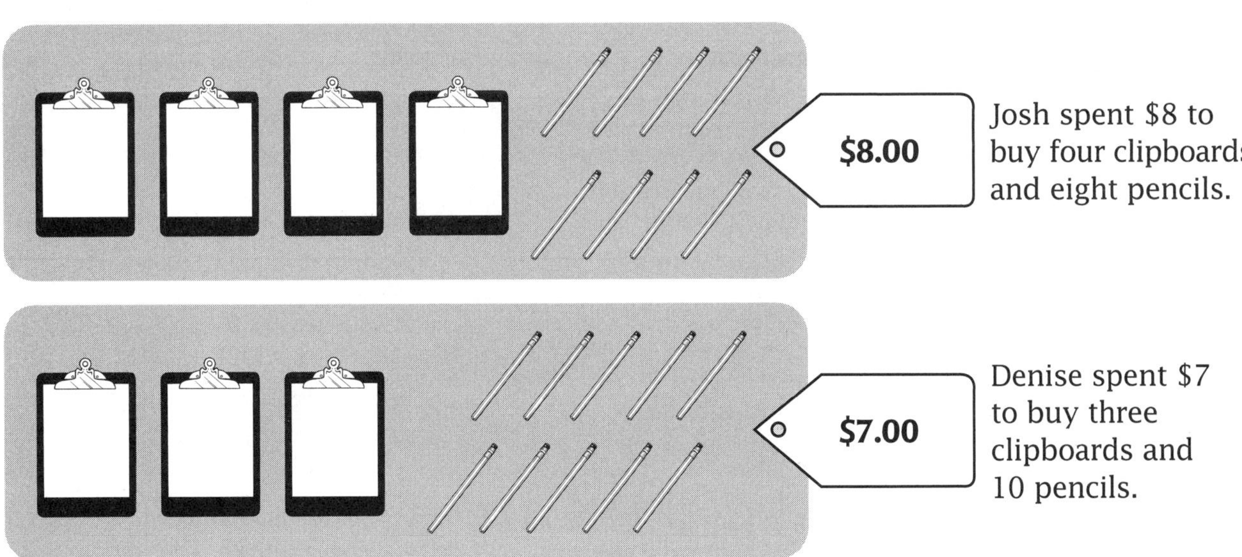

Josh spent $8 to buy four clipboards and eight pencils.

Denise spent $7 to buy three clipboards and 10 pencils.

Denise wants to trade Josh two pencils for a clipboard.

9. Is that a fair exchange? If not, who has to pay the difference, and how much is it?

10. What is the price of a pencil? What is the price of a clipboard? Explain your reasoning.

Costs of Combinations (in dollars)

Number of Umbrellas						
5						
4						
3						
2		80				
1			76			
0						
	0	1	2	3	4	5

Number of Caps

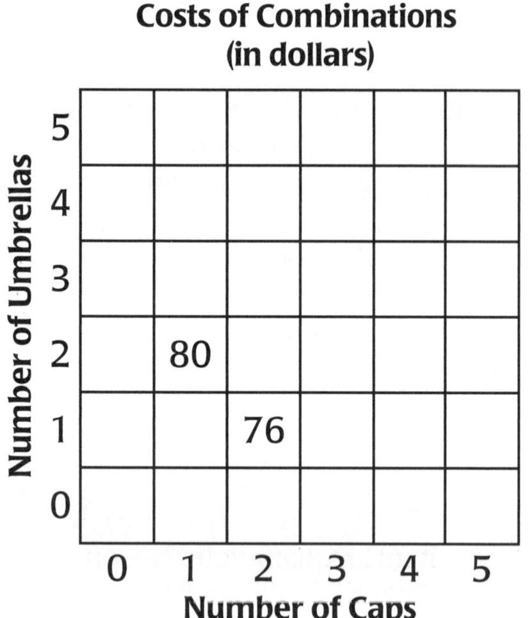

Some of these shopping problems can be solved using a chart.

On the left you see a part of the combination chart that fits the problem of the caps and the umbrellas (page 16).

11. Fill in this chart on **Student Activity Sheet 4.** Then find the prices of one cap and one umbrella. Is this the same answer you got for problem **7** on page 16?

12. Look at the two pictures of eyeglasses and shorts on page 15. Use one of the extra charts on **Student Activity Sheet 4** to make a combination chart for these items. Label your chart. What is the price of one pair of glasses? one pair of shorts?

Use any strategy you wish to solve problems **13** and **14.**

At Doug's Discount Store, all compact discs are one price; all videotapes are another price.

David buys three CDs and two videotapes for $67.

Joyce buys two CDs and four videotapes for $90.

13. What is the price of one CD? one videotape?

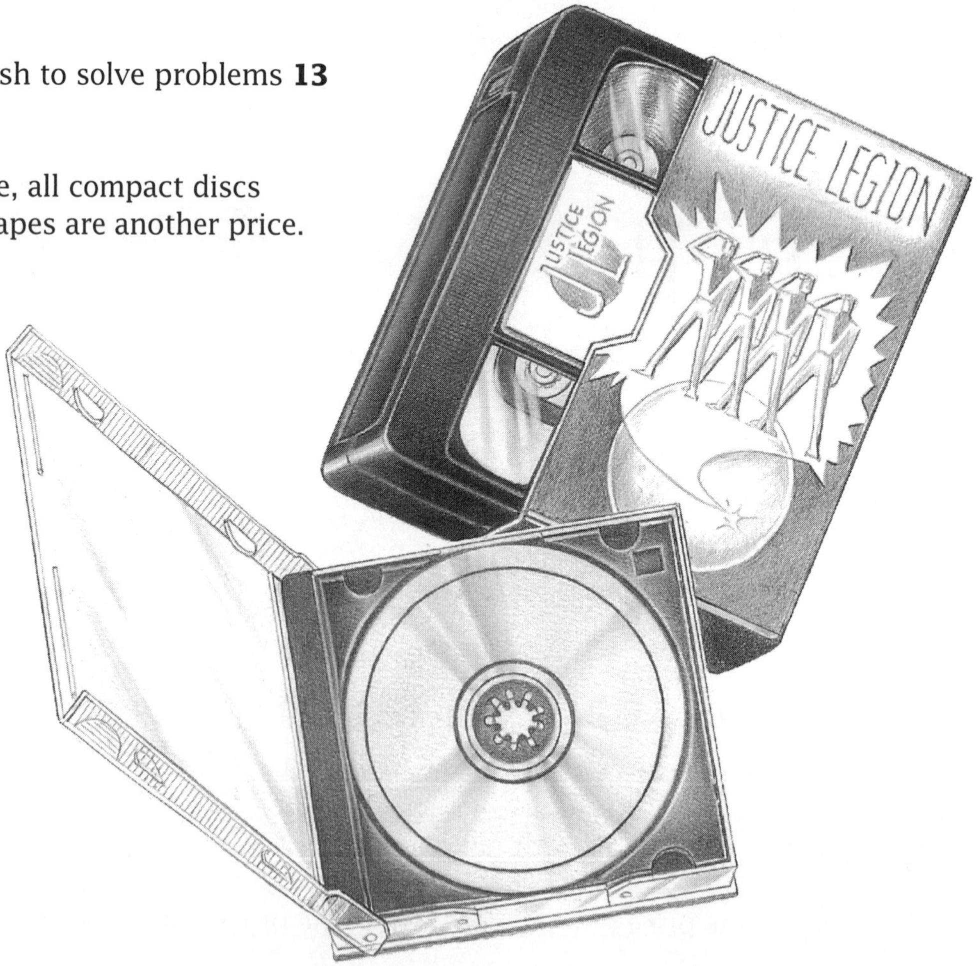

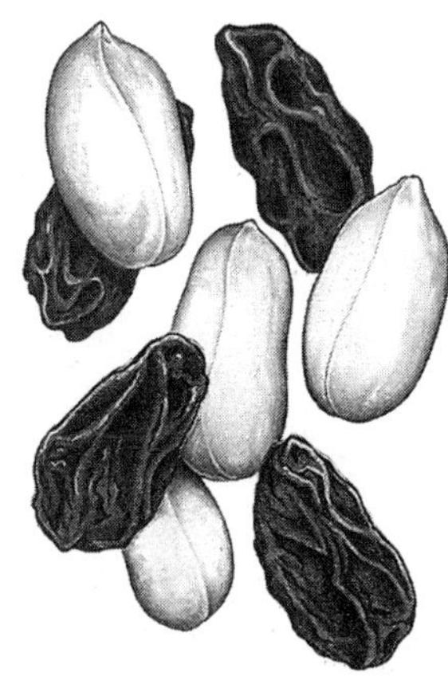

On a visit to Quinn's Quantities, Rashard finds the following prices for various combinations of peanuts and raisins:

- A mixture of 3 cups of peanuts and 2 cups of raisins costs $3.30.

- A mixture of 4 cups of peanuts and 3 cups of raisins costs $4.55.

14. What does Rashard pay for a mixture of 5 cups of peanuts and 2 cups of raisins?

15. Create your own shopping problem. Be sure you can solve the problem, then ask someone else to solve it. Have the person explain to you how he or she found the solution.

In solving shopping problems, you have used exchanging and combination charts. Joe looked at the problem below and used a different strategy.

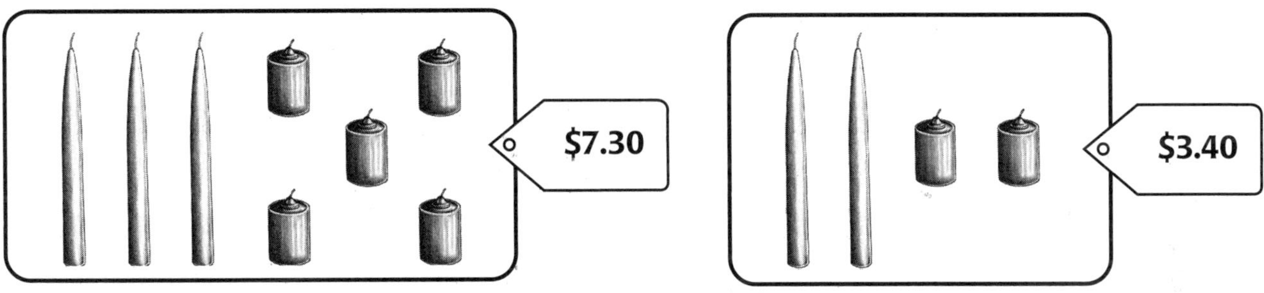

Below you see how Joe found the price of each candle.

16. Explain Joe's reasoning.

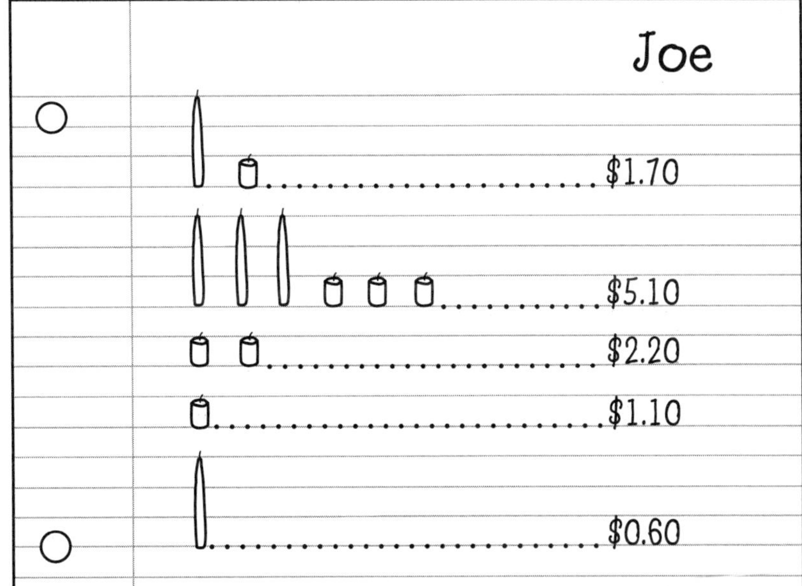

Summary

Shopping problems can be solved using the method of exchange. Identifying a pattern in a picture or combination chart can make it possible to find the cost of one item. Extending a pattern or combining information can also help you find the cost of one item.

Summary Questions

Suppose the prices of the candles changed.

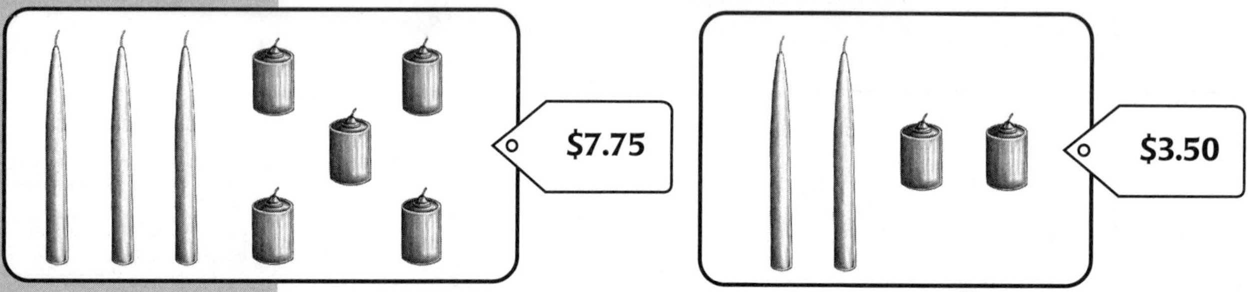

17. Margarita tried to find the price of each candle with a combination chart. Use one of the extra charts on **Student Activity Sheet 4** to show how Margarita might have solved the problem.

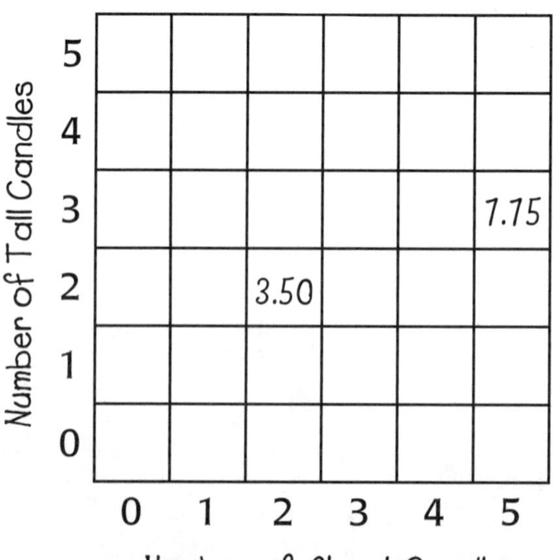

Chickens

Three chickens weighed themselves in different groupings.

1. What should the scale read in the fourth picture?

2. Now you can find out how many kilograms each chicken weighs. Show how.

Mario's Restaurant

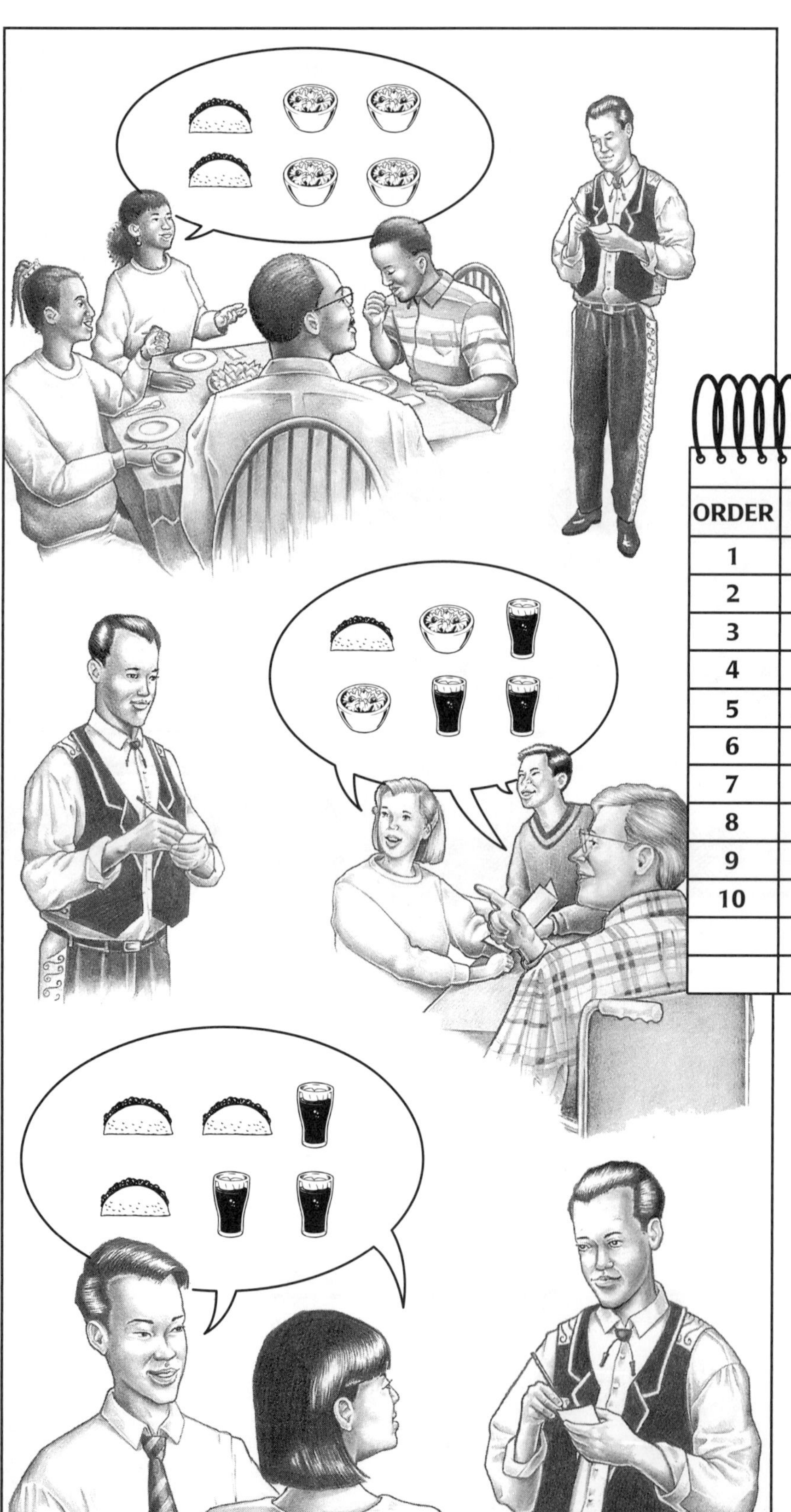

Mario runs a Mexican restaurant, and he is very busy. He moves from one table to another, writing down all the orders. Below you see how he writes the orders on his order pad.

3. Some of the orders do not have total prices indicated. What are the prices of these orders?

4. Make up two new orders and write them in your notebook. Fill in the prices of these orders.

5. What is the price of each individual item?

Chickens Revisited

The way Mario wrote the orders in his notebook gave him a good overview of many combinations. Such notation can also be used for other problems. If you use Mario's *notebook notation* for the chicken problem, you might get the following:

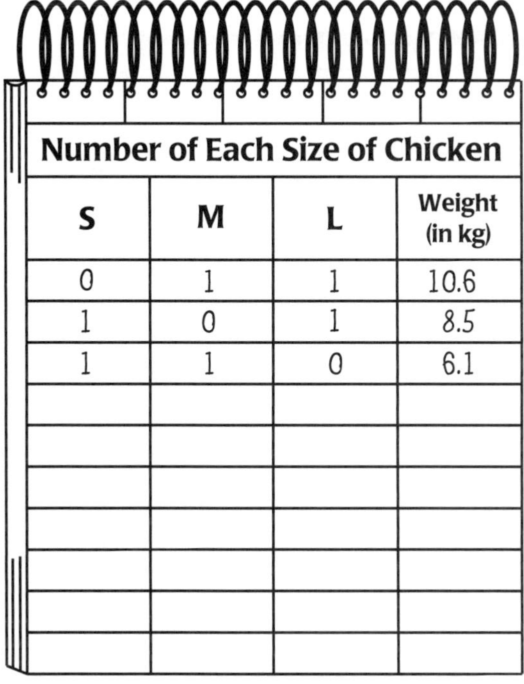

Number of Each Size of Chicken

S	M	L	Weight (in kg)
0	1	1	10.6
1	0	1	8.5
1	1	0	6.1

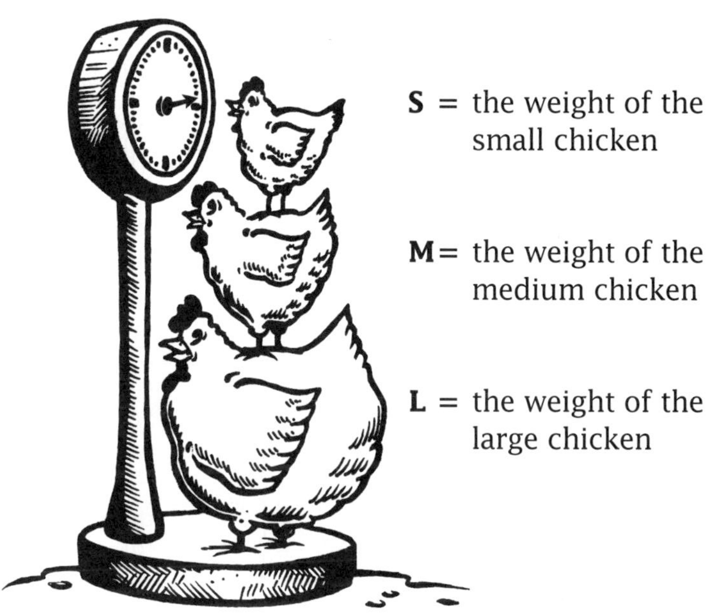

S = the weight of the small chicken

M = the weight of the medium chicken

L = the weight of the large chicken

6. How can you find the total weight of the three chickens using notebook notation?

7. Make new combinations until you find the weight of each chicken.

Burger World

To the right are some orders that were served at Burger World today. You can write these orders in a notebook, as shown below.

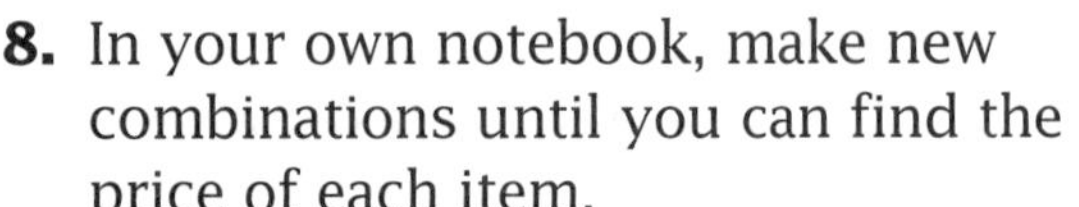

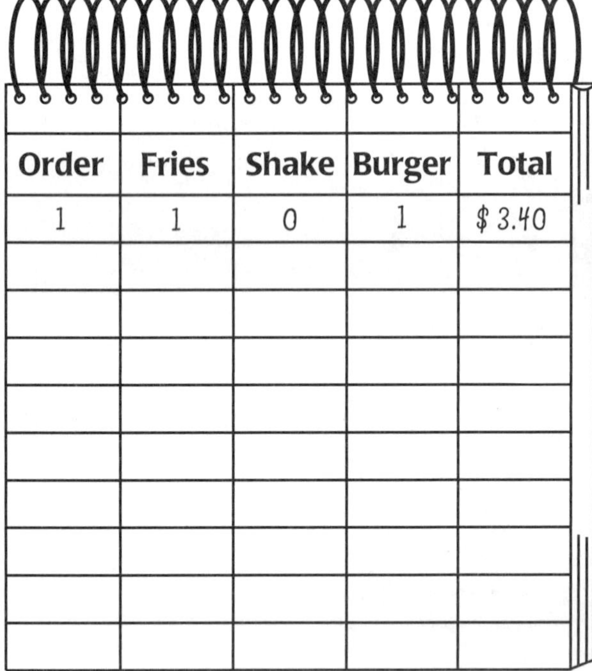

Order	Fries	Shake	Burger	Total
1	1	0	1	$ 3.40

8. In your own notebook, make new combinations until you can find the price of each item.

Bouquet of Flowers

Mandy wants to buy some flowers. At a flower stand, she sees some beautiful bouquets at different prices.

For $6.20, she can buy a bouquet of two lilies and six chrysanthemums.

For $8.60, she can buy a bouquet of six roses and four chrysanthemums.

For $5.00, she can buy a bouquet of four roses and two lilies.

Mandy cannot decide which bouquet to buy because she likes all the flowers.

She asks the florist to make up a bouquet of lilies, roses, and chrysanthemums that will cost $10 or less.

9. Find one possibility for Mandy's bouquet.

Another customer wants a bouquet of 24 flowers.

10. Find three combinations of 24 flowers. Calculate the price of each bouquet. Record your calculations in your notebook.

Summary

In this section, you have seen that notebook notation is a good way to get an overview of the information contained in a problem. You can make new combinations in a notebook by

- adding rows,

- taking the difference between rows,

- doubling or halving rows, and so on.

The new combinations you create can help you find a solution to a given problem.

Summary Questions

Solve the following problem using notebook notation. Show all of your calculations.

For a total cost of $18.40, Gideon went on the Whirling Wheel four times, in the Haunted House two times, and on the Roller Coaster four times.

For a total cost of $18, Louisa went on the Whirling Wheel five times and on the Roller Coaster five times.

Bryce likes only the Roller Coaster, and he rode it 10 times! He spent one dollar less than Louisa.

11. What is the price of each attraction? Show all of your calculations.

12. Create a problem of your own using notebook notation. Show a detailed solution to your problem.

The School Store Revisted

The prices of pencils and erasers have changed, so Martin and Monica have to make a new price chart. **Student Activity Sheet 5** contains a combination chart for you to write on.

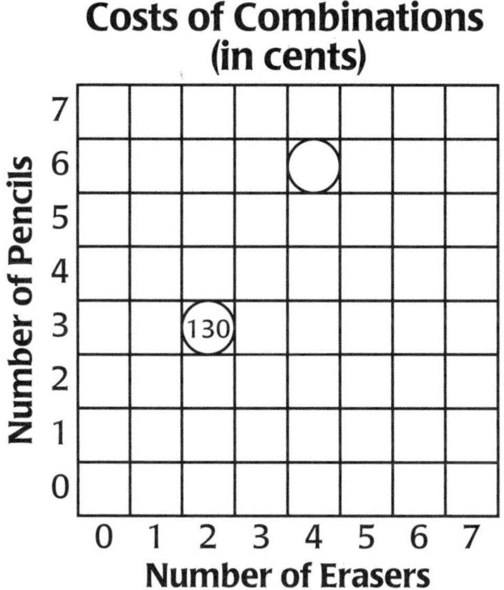

Information about the number in the circle can be expressed in a formula:

$$2E + 3P = 130$$

Such a formula is called an *equation*.

1. Describe in words the meaning of this equation.

2. What number should go in the empty circle above? Write an equation representing this entry.

3. Use **Student Activity Sheet 5** to write the information from the two equations in notebook notation.

4. Monica tells you that $1E + 2P = 75$. Put this information in both the combination chart and the notebook on **Student Activity Sheet 5.**

5. Find the new prices of one eraser and one pencil. You may use notebook notation or the combination chart.

Hats *and Glasses*

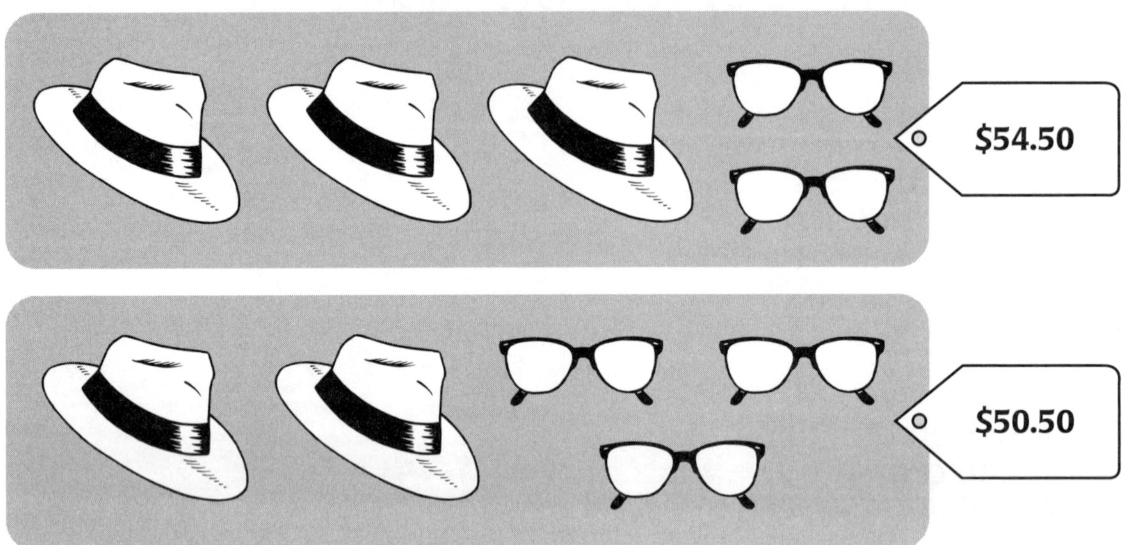

6. Each of the pictures above can be replaced by an equation. Write the two equations using the symbol *H* for the price of a hat and the symbol *G* for the price of a pair of glasses.

7. Write an equation that shows the total price of one hat and four pairs of glasses.

8. What is the price of one pair of glasses? What is the price of one hat? Show how you found these prices.

Below is another chart for the costs of combinations of two items.

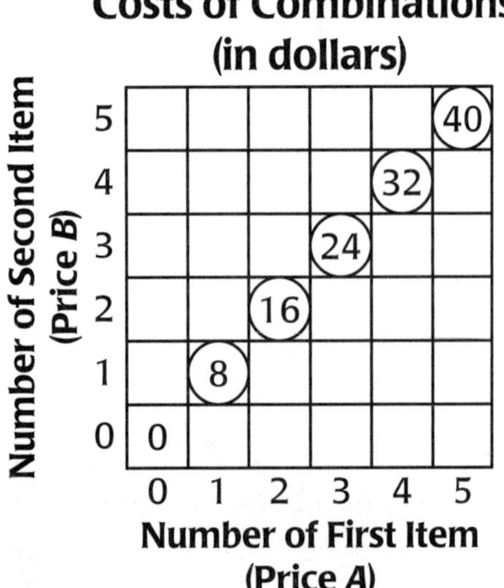

9. Write an equation in which the price of one item is *A* and the price of the other item is *B* for each of the circled numbers. You should have five different equations.

10. Make up a price for each item so that price *A* is higher than price *B*. Use these prices to complete the chart in your notebook.

11. Do you think all of the students in your class have the same numbers in their charts? Explain why or why not.

12. Now you can make many equations corresponding to your chart. Write three of them.

Return to Mario's

Some prices in Mario's restaurant have changed. You now have to pay $6.50 if you order one taco, two salads, and one drink. You have to pay $11.50 for one taco, four salads, and three drinks. For $4.50 you can get one taco and two drinks.

13. Write an equation that corresponds to each of the orders above.

14. By combining the orders, you can make new equations. What equation do you get when you add the last two orders?

15. Make up two other equations by combining orders.

16. Show how you can combine equations to get the equation $1S + 1D = \$2.50$.

17. Find the new price for each of the three items.

This afternoon, there is a new animated movie playing at the movie theater. Many adults and children are waiting to buy their tickets.

18. What will the ticket seller in the third picture say?

19. How much would you have to pay if you went to this theater alone?

Summary

Many problems compare quantities such as prices, weights, or widths.

One way to describe these problems is by using *equations.* Look at the picture below, for example.

If you let U represent the price of one umbrella and C represent the price of one cap, the equation would be $2U + 1C = \$80$.

These types of problems can also be solved with combination charts or notebook notation.

Costs of Combinations

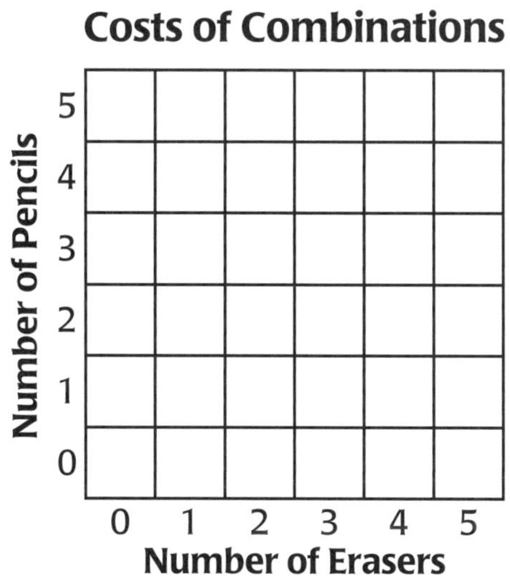

Notebook Notation

Summary Questions

Here are two equations:

$$4L + 3M = 96$$

$$L + M = 27$$

The numbers 96 and 27 can represent lengths, weights, prices, or whatever you wish.

20. Find the value of L and the value of M.

21. Make up a story to fit these equations.

Section A. Compare and Exchange

Susan and her friends like to collect and trade basketball cards. Today after school, Susan made the following trades:

- two Slamming Sams for three Halima Hoopsters,
- three Fast Break Freddys for four Slamming Sams,
- one Fast Break Freddy for one Slamming Sam and two Rebound Ritas,
- four Jumping Jeans for two Fast Break Freddys.

1. Use the above information to make up two more fair card trades.

2. James offers Susan six Halima Hoopsters for three Fast Break Freddys. Should Susan make this trade? Why or why not?

3. James then offers one Jumping Jean for two Slamming Sams. Should Susan make this trade? Why or why not?

4. Susan has five Fast Break Freddys. How many Rebound Ritas can she get for her Fast Break Freddys?

Section B. Looking at Combinations

There are going to be two rides at this year's school fair. The Loop-D-Loop requires five tickets for one ride, and the Whirlybird requires two tickets.

1. Copy in your notebook the following combination chart that shows how many tickets are needed for different combinations of these two rides. Fill in the chart as necessary to solve the following problems.

2. How many tickets are needed for two Loop-D-Loop and three Whirlybird rides?

3. Janus has 19 tickets. How can she go on both rides and have no leftover tickets?

4. a. Mark on your combination chart a move from one square to another that represents the exchange of one Whirlybird ride for two Loop-D-Loop rides.

   b. By how much does the number of tickets change?

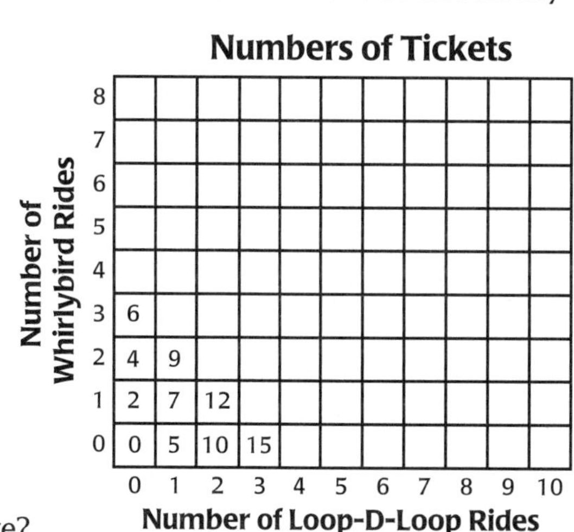

5. For each of the following puzzles, find the number that goes in the circled box and explain your strategy.

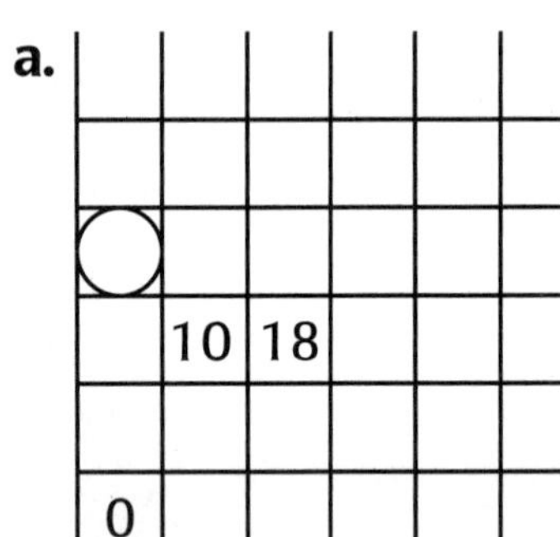

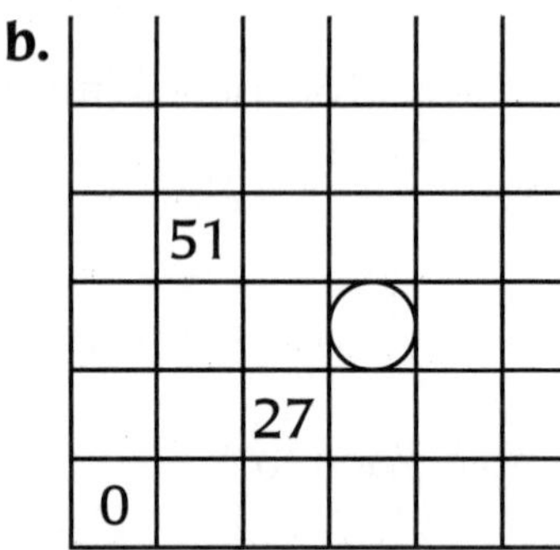

Section C. Finding Prices

1. Felicia and Kenji bought the following candles.

    a. Without calculating the price for a short or tall candle, determine which is more expensive. How much more expensive is it?

    b. Use the two pictures to make a new combination of short and tall candles. Write the cost of the combination.

    c. What is the price of one short candle? one tall candle?

2. Roberto bought three T-shirts and four caps for $96. Anne bought two T-shirts and five caps for $99.

    a. Make a combination chart to represent this information. Be sure to label the horizontal and vertical sides of your chart.

    b. What is the price of one T-shirt? one cap?

Section D. Notebook Notation

1. Some of today's orders at Fish King are shown in the notebook below.

ORDER	DRINK	FRIES	FISH	TOTAL
1	2	1	2	$ 8.80
2	1	--	1	$ 3.60
3	3	1	1	$7.40
4				
5				
6				
7				

   a. In your own notebook, list at least three new combinations.

   b. What is the price of each item at Fish King?

2. Study the following notebook showing lunch orders at Gino's restaurant.

ORDER	TACO	SALAD	DRINK	TOTAL
1	1	--	2	$ 3.00
2	2	1	4	$ 8.00
3	--	4	4	$ 11.00
4				
5				
6				
7				

   a. Find the cost of one salad. Explain how you got your answer.

   b. How can you find the cost of one drink? one taco?

3. Can you solve problems **1** and/or **2** using combination charts? Why or why not?

Section E. Equations

1. At a flower shop, Joel paid $10 for three irises and four daisies. Althea paid $9 for two irises and five daisies.

    a. Write equations representing this information.

    b. Write an equation that shows the price of one iris and six daisies.

    c. Find the cost of one iris. Find the cost of one daisy.

2. At a movie theater, tickets for three adults, two seniors, and two children cost $35. Tickets for one senior and two children cost $12.50. Tickets for one adult, one senior, and two children cost $18.50.

    a. Write three equations representing the above information about the movie theater. Use A to represent the price of an adult's ticket, S to represent the price of a senior's ticket, and C to represent the price of a child's ticket.

    b. Make up two other equations by combining your first three equations.

    c. Explain how you can combine equations to get the equation $2A + 1S = \$16.50$.

    d. Explain how you can combine equations to get the equation $A = \$6$.

    e. What is the cost of each ticket?

3. Here you see two equations:

    $$8C + 3K = 46$$
    $$5C + 3K = 40$$

    Find the value of C and the value of K.

Cover

Design by Ralph Paquet/Encyclopædia Britannica, Inc.

Collage by Koorosh Jamalpur/KJ Graphics.

Title Page

Illustration by Phil Geib/Encyclopædia Britannica, Inc.

Illustrations

1, **2**, **3 (top)** Phil Geib/Encyclopædia Britannica, Inc.; **3 (bottom)** Paul Tucker/Encyclopædia Britannica, Inc.; **4** Phil Geib/Encyclopædia Britannica, Inc.; **5** Paul Tucker/Encyclopædia Britannica, Inc.; **6–7, 9, 11–13, 15–19, 21** Phil Geib/Encyclopædia Britannica, Inc.; **22** Paul Tucker and Phil Geib/ Encyclopædia Britannica, Inc.; **23–25, 27–28, 30, 33** Phil Geib/ Encyclopædia Britannica, Inc.

Photographs

15 © Robert E. Daemmrich/Tony Stone Images; **24** © Billy E. Barnes/Tony Stone Images; **25, 29** © Encyclopædia Britannica, Inc.